meet the mammoth!

The Fossil Files

pamela pettyfeather

PETTYFEATHER
PUBLISHING

contents

introduction: what's a woolly mammoth?

Picture this: an elephant with fur thicker than a winter coat and tusks longer than a car.

No joke. It had a built-in parka, snow-proof feet, and curving ivory tusks that could shovel through ice and snow. And it wasn't just surviving in the cold—it was ruling the Ice Age.

Meet the woolly mammoth.

That name might sound made-up, like something from a cartoon, but woolly mammoths were very real. They stomped across Europe, Asia, and North America during the last Ice Age, from around 400,000 years ago to just a few thousand years ago. They shared the Earth with saber-toothed cats, cave lions, dire wolves—and early humans. People hunted them, painted them on cave walls, and built shelters from their bones. Some humans even lived alongside them.

Here's the really amazing part: In some places, mammoths didn't fossilize or rot away. They froze solid.

Scientists have found frozen mammoth bodies buried in Arctic ice, complete with fur, skin, and even their last meals still in their bellies. These frozen giants give us clues about mammoth life, mammoth death, and maybe, just maybe, mammoth resurrection.

But first, let's discover what made the woolly mammoth one of the coolest creatures of all time.

1
giant, hairy, and built for cold

LET'S get one thing straight. The woolly mammoth was not just an elephant wearing a sweater.

It was a cold-weather champion designed from trunk to tail to handle freezing winds, deep snow, and icy ground. While other animals shivered and struggled, the mammoth marched through blizzards like it was a sunny day.

Here's what made this chilly champion so special.

Built-In Winter Gear

The woolly mammoth had serious cold-weather upgrades.

Body Part	Ice Age Advantage
Fur	A double-layered coat with long, shaggy outer hair and soft, warm underfur
Fat	Up to 4 inches of fat under the skin for insulation
Ears & Tail	Short and stubby—less surface area to lose heat
Feet	Wide, padded soles for walking on snow like built-in snowshoes
Back Hump	A fatty hump to store energy, like a survival backpack

This wasn't just an elephant in disguise. It was a cold-adapted tank.

body part: ice age advantage

Fur: A double-layered coat with long, shaggy outer hair and soft, warm underfur

Fat: Up to 4 inches of fat under the skin for insulation

Ears & Tail: Short and stubby—less surface area to lose heat

Feet: Wide, padded soles for walking on snow like built-in snowshoes

Back Hump: A fatty hump to store energy, like a survival backpack

This wasn't just an elephant in disguise. It was a cold-adapted tank.

tusks that told stories

Woolly mammoth tusks were more than decorations. They were tools, weapons, snow shovels, and social signals all in one.

Some tusks grew over fifteen feet long and curved in spirals. Mammoths used them to:

- Dig through snow to reach buried grass
- Defend themselves against predators
- Fight with other mammoths over territory or mates
- Show off their strength and age

Older mammoths often had tusks with cracks, chips, and scrapes—battle scars from a tough life in the wild.

elephant cousins... but not twins

WOOLLY MAMMOHS WERE RELATED TO TODAY'S ELEPHANTS, BUT THEY WEREN'T IDENTICAL.

Feature	Woolly Mammoth	Asian Elephant	African Elephant
Height	Up to 13 feet	Up to 10 feet	Up to 13 feet
Hair	Thick, shaggy fur	Sparse hair	Barely any
Tusks	Long & curved	Shorter & straighter	Long, slightly curved
Ears	Small & rounded	Medium-sized	Very large
Habitat	Cold grasslands & tundra	Forests	Savannas & woodlands

They're all part of the elephant family tree—but the mammoth took a frosty detour.

herd life on ice

Mammoths didn't wander alone. They traveled in **herds**, just like elephants do today. A typical herd included:

- A matriarch (the oldest female, the boss)
- Other adult females
- Young mammoths and calves

Males usually left the herd when they got older and roamed solo or in bachelor groups.

Herds helped protect young calves, find food, and remember safe paths through the wild Ice Age world. Scientists think mammoths were smart and social—using their trunks to touch, explore, and comfort each other.

Brain Boost: The Mammoth's Trunk

A woolly mammoth's trunk was like a super-tool. It didn't have bones inside—it was made of thousands of tiny muscle bundles all working together, like how a rope is made of many smaller threads twisted together. These muscle bundles could squeeze, stretch, and twist to make the trunk move in amazing ways!

Mammoths used their trunks to:

- Smell and sniff out food
- Grab grasses or break twigs
- Spray water or dust
- Make sounds by trumpeting
- Greet other mammoths with trunk hugs

That's more flexible than a robot arm—and much cuter.

try this: *draw your own mammoth*

Design your very own woolly mammoth. Label the parts that helped it survive:

- Long tusks
- Shaggy fur
- Small ears
- Big feet
- Fatty back hump
- Trunk (don't forget the wrinkles!)

Then imagine it marching through the snow hunting for grass.

2
life on the mammoth steppe

TIME TO STEP BACK—WAY back—into the Ice Age.

No roads. No buildings. No farms. Just endless frozen plains stretching to the horizon. The air stings your nose. The wind howls. And in the distance, a herd of woolly mammoths trudges across the snowy ground.

This was the mammoth's world.

welcome to the mammoth steppe

The mammoth steppe was one of the largest ecosystems in Earth's history. It stretched from western Europe across Russia, through Siberia, and into Alaska and Canada. Think of it as the Ice Age version of a superhighway—wide, flat, and covered in grass, not trees.

Despite the cold, this land teemed with life. Grasses, mosses, herbs, and shrubs carpeted the ground, making it perfect for giant plant-eaters like the woolly mammoth.

For a mammoth, this was paradise.

brain boost: land bridges and ice roads

During the Ice Age, sea levels were much lower. Places that are separated today—like Russia and Alaska—were connected by land. Scientists call this area **Beringia**.

Thanks to that frozen bridge, mammoths could walk between continents without boats, planes, or passports.

This helped them spread from Europe to the edges of North America.

what's for dinner?

Woolly mammoths were plant-eaters through and through. With their huge molars, they could grind tough vegetation like walking lawnmowers. And they ate constantly. A full-grown mammoth munched over **300 pounds of plants** every day.

That's like eating a kiddie pool full of salad—for breakfast alone.

Mammoth Munchies

Grasses and sedges

Shrubs and twigs

Mosses and herbs

Bark (when nothing else was around)

meet the ice age neighbors

The mammoth steppe wasn't just home to mammoths. It was packed with giant, furry, and sometimes frightening creatures:

- **Woolly Rhinoceros**: Another shaggy tank with two big horns
- **Cave Lion**: A massive predator, bigger than today's lions
- **Saiga Antelope**: A strange grassland grazer with a puffy nose
- **Steppe Bison**: Fast, powerful, shaped like a bulldozer with horns

- **Arctic Foxes and Wolves**: Smaller but speedy scavengers
- **Giant Deer**: With antlers wider than a car
- **Humans**: Tool-making, fire-building hunters exploring new worlds

It was like a prehistoric safari, but with snow instead of sand.

where did mammoths live?

Woolly mammoths weren't picky. As long as there was food, they could handle almost anything.

They roamed:

- Snowy plains and frozen tundras
- Grassy valleys with icy rivers
- Windy hills and open steppe

Fossils have been found across:

- Europe
- Russia and Siberia
- Mongolia and northern China
- Alaska
- Canada
- The northern United States

They even reached islands off the coast of Siberia—and some mammoths stayed there long after the others disappeared.

life in a tough world

Even for a mammoth, Ice Age life was challenging. Dangers lurked everywhere:

- Sudden blizzards and harsh winters
- Thin ice and deep snow
- Packs of hungry predators
- Long migrations searching for food

But mammoths had teamwork on their side. They lived in herds that looked out for each other. Calves stayed protected by adults. Mothers guided the group. Together, they marched across the coldest places on Earth.

try this: *map the mammoth steppe*

Find a world map and mark these places:

- Siberia
- Alaska
- Canada
- Northern Europe

Draw a giant grassy zone connecting them. That's the mammoth steppe.

Add symbols to show where different Ice Age animals lived. Draw mammoths, rhinos, lions, bison—even early humans. Create a map of the wildest winter world ever.

3
tusks, trunks, and herd life

IF YOU WERE A WOOLLY MAMMOTH, you wouldn't spend your days wandering alone through the snow.

You'd be part of a team—a tight-knit herd of mothers, calves, sisters, and aunties. You'd follow your matriarch through the frozen wilderness, munching grass, watching for predators, and learning to survive one chilly challenge at a time.

Let's see how mammoths lived, moved, and thrived—together.

meet the mammoth herd

Woolly mammoths were social animals, just like elephants today. Their herds worked like elephant families:

- Led by a **matriarch**—the oldest, wisest female
- Filled with **related females and their calves**
- Sometimes included **young males**, until they left the group

The herd moved together, foraged together, and even grieved when one of their own died. Fossils show mammoths often traveled in groups and returned to the same places year after year— just like migrating animals today.

trunks: the mammoth multi-tool

A mammoth's trunk wasn't just a nose—it was a super-tool with thousands of muscle bundles.

Mammoths used their trunks to:

- Sniff out food buried under snow
- Tear up grass and twigs
- Slurp water from icy streams
- Trumpet warnings or greetings
- Touch and comfort other mammoths

Their trunks were strong enough to lift heavy branches and precise enough to pluck a single blade of grass.

Scientists think mammoths could "talk" with their trunks—using touches, rumbles, and movements to communicate in ways we're still trying to understand.

tusks that did everything

Woolly mammoths had some of history's most impressive tusks. These tusks:

- Could grow **12 to 15 feet long**
- Curved in spirals from their faces
- Grew throughout their lives—like very slow fingernails

Mammoths used their tusks for many jobs:

- **Shoveling snow** to reach buried food
- **Breaking ice** to get to water
- **Fighting rivals** during mating season
- **Defending** against predators
- **Showing status** (older bulls had bigger tusks)

Some tusks even have **growth rings**, like trees. By studying them, scientists can tell what a mammoth experienced each year —whether it was healthy, stressed, or even pregnant.

Trunk vs. Tusk

People sometimes mix these up.
Here's how to tell them apart.

	What It Is	What It Does
Trunk	Long, flexible nose	Smelling, grabbing, drinking, trumpeting
Tusks	Long, curved teeth	Digging, fighting, lifting, showing off

Trunks = nose-tools. Tusks = mega-teeth.

the power of the herd

Being in a herd gave mammoths major advantages:

- **Protection** — More eyes to watch for danger
- **Learning** — Young calves copied the adults
- **Memory** — Older females remembered migration routes and safe spots
- **Care** — If a calf got stuck or sick, the herd helped

Fossils even show **circles of mammoths** that may have been protecting a fallen calf—or mourning a loss.

That's not just survival. That's community.

what about the bulls?

Adult male mammoths usually left the herd around age 10-15. After that, they wandered solo or formed **bachelor groups**. During mating season, bulls would return to find a mate—but they didn't stay long.

These males were often larger with thicker tusks, and sometimes got into serious **tusk-to-tusk battles** over mates.

Fossil skulls with broken tusks and cracked jaws? Probably the result of an intense Ice Age showdown.

try this: build your mammoth herd

Create a "family tree" for a woolly mammoth herd. Include:

- A matriarch (name her!)
- Her sisters or daughters
- A few calves
- A young male getting ready to leave

Draw each mammoth, give them names, and describe their job in the herd. Who finds food? Who protects the calves? Who remembers the trail?

4
mammoths and humans

LONG BEFORE CITIES, cars, or computers, humans lived in small bands, traveled on foot, and used stone tools to survive.

And one of the most incredible animals they ever met? The woolly mammoth.

Our Ice Age ancestors didn't just see mammoths—they **lived alongside them**, hunted them, painted them, and told stories about them. Mammoths weren't just animals to early humans. They were neighbors, resources, and legends.

mammoth hunters

You might think it would take a superhero to bring down a mammoth. But Ice Age humans did it—with teamwork, clever tools, and courage.

Here's how they likely hunted:

- Tracked herds across open plains
- Targeted weak, young, or injured mammoths
- Used **spears, traps, or cliffs** to slow them down
- Worked in groups to surround and attack

It was dangerous. A charging mammoth could weigh six tons and run faster than most humans. But one successful hunt could feed a whole group for weeks—and provide tools, shelter, and clothes too.

More Than Meat

When humans hunted a mammoth, they used everything.

Body Part	How Humans Used It
Meat	Food for days or weeks
Fat	Cooking fuel and insulation
Tusks	Tools, weapons, carvings
Bones	Building shelters, making furniture
Hide	Blankets, tents, clothing

In some Ice Age camps, archaeologists have found **entire houses** built from mammoth bones and covered in hides. Imagine living inside a mammoth ribcage with a fire pit in the middle!

art and memory

Early humans didn't just use mammoths—they **remembered** them. They painted mammoths on cave walls, carved them into tools, and sculpted tiny mammoth figurines from ivory.

Some of these artworks are **over 30,000 years old**.

Cave art has been found in:

- Lascaux and Chauvet (France)
- Kapova Cave (Russia)
- Vogelherd (Germany)

These ancient paintings show mammoths in motion—trumpeting, marching, or standing guard. They weren't just decorations. They were part of human memory, storytelling, and spiritual life.

Brain Boost: The Mammoth in Our Myths

Some scientists think mammoth bones might have inspired ancient legends:

- A huge skull with a hole in the middle? Ancient people might have thought it was a **cyclops**.
- A giant curved tusk? Maybe it became the horn of a **sea serpent**.
- A frozen baby mammoth? That's a story waiting to be told.

Even today, mammoths appear in cartoons, museums, video

games, and books. Once they walked beside us—now they live in our imaginations.

try this: *make your own mammoth art*

Design a cave wall scene with woolly mammoths and early humans.

Use only **earth tones** like red, brown, and black—just like ancient cave artists did.

Include:

- A mammoth herd on the move
- Hunters with spears
- Firelight dancing on the walls

Give your art a title. What story does it tell?

5
the great mammoth disappearance

FOR HUNDREDS of thousands of years, woolly mammoths roamed the frozen north—marching through snow, trumpeting across valleys, raising their young, and surviving Ice Age after Ice Age.

And then... they were gone.

No more mammoth herds. No more trumpeting calls. Just silence—and bones buried deep beneath the Earth.

What happened? Let's solve one of the Ice Age's biggest mysteries.

the world warms up

About 12,000 years ago, Earth's climate began changing. The last Ice Age was ending, and the planet was getting warmer:

- Glaciers melted and retreated
- Forests spread into areas that were once grassland
- The mammoth steppe—the wide, frozen grassland they called home—began disappearing

This was bad news for mammoths. Their favorite food (cold-climate grasses) vanished. Their stomping grounds turned into soggy forests. And their thick fur and heavy bodies weren't built for warm, wet places.

They didn't go extinct overnight. But over thousands of years, mammoths faced shrinking space, fewer resources, and tougher competition.

humans on the rise

At the same time climate was changing, something else was happening: humans were spreading. From Asia to Europe to North America, early humans were exploring new lands—and crossing paths with mammoths more often.

They had:

- **New hunting tools** like spears and spear-throwers

- **Fire** to clear land and stay warm
- **Cooperation and communication** to plan hunts
- **Growing populations** that needed food and shelter

Humans didn't wipe out mammoths instantly—but their hunting may have added pressure when mammoths were already struggling.

Sometimes, one extra challenge is all it takes.

the last mammoths

Most mammoths were extinct by about 10,000 years ago. But not all of them.

Incredibly, a small group survived on **Wrangel Island**, off the coast of Siberia, until about **4,000 years ago**—that's **after** the pyramids in Egypt were built.

These island mammoths were smaller, possibly due to limited resources, and may have suffered from inbreeding and poor genetic health.

Eventually, they disappeared too. When the last Wrangel Island mammoth died, the world's most iconic Ice Age animal was gone forever.

What Causes Extinction?

Cause	Example
Climate Change	Ice Age ends, mammoth habitat disappears
Loss of Food	Plants and grazing land vanish
Overhunting	Humans hunt faster than animals can reproduce
Habitat Loss	Forests replace grasslands
Genetic Problems	Small populations can't stay healthy

It's often not just one thing—it's a combination.
Scientists call this a "perfect storm."

try this: *mammoth mystery timeline*

Make a timeline showing:

- When mammoths first appeared (about 400,000 years ago)
- When humans began to spread
- When the last Ice Age ended
- When most mammoths went extinct
- When the last island mammoths vanished

Mark where climate and humans both played a role. What patterns do you notice?

6
frozen in time

Lyuba, the Baby Mammoth*

WOOLLY MAMMOTHS MAY BE EXTINCT—BUT they
left behind one of the **coolest fossil records** in history. Literally.

Unlike most ancient animals, mammoths didn't always rot away or turn to stone. In the coldest corners of Earth, some froze solid —preserved for thousands of years in ice, snow, and frozen ground.

That means we've found real mammoth skin, real mammoth hair, even real mammoth **stomachs**.

Let's see what the ice kept hidden—and what it's revealing today.

ice age time capsules

When an animal dies, it usually disappears quickly. The body breaks down, gets eaten, or decays. But in places like **Siberia, Alaska, and the Yukon**, there's something different: **permafrost**.

Permafrost is ground that stays frozen all year round. In this natural deep freezer, woolly mammoths were preserved in amazing detail.

Some discoveries include:

- Skin and muscle tissue
- Long reddish-brown fur
- Stomach contents (grasses and flowers!)
- Teeth and tusks in perfect condition
- Even tiny baby mammoths, curled up as if asleep

meet lyuba, the baby mammoth

In 2007, a reindeer herder in Russia made one of the most incredible Ice Age discoveries ever: a perfectly preserved baby woolly mammoth.

She was named **Lyuba**.

She had:

- Her full coat of hair
- Wrinkled skin
- All her organs
- Milk in her belly
- Even eyelashes

Lyuba was only about **one month old** when she died—likely trapped in mud and frozen soon after. She's been called the best-preserved mammoth ever found.

Thanks to her, scientists learned more about mammoth parenting, diet, and how baby mammoths grew.

mammoth gold rush?

As the world warms, more frozen ground is melting—especially in Siberia. That means more mammoth tusks and bones are appearing... and some people are collecting and **selling** them.

- Mammoth ivory is legal in many places (unlike elephant ivory)
- Some say this helps protect modern elephants
- Others say it encourages dangerous trade and damages important fossils

It's like a modern-day **mammoth gold rush**—and it raises big questions about how we treat ancient remains.

What Frozen Mammoths Tell Us

Frozen mammoths are like nature's notebooks—full of clues from the past.

Body Part	What It Tells Us
Hair & skin	What color mammoths were; how they kept warm
Teeth & tusks	How old they were, what they ate
Stomach contents	What plants grew nearby
Bones & joints	How they moved, grew, and aged
Back Hump	How mammoths are related to elephants & each other

With every frozen find, we learn more about how mammoths lived... and why they vanished.

try this: *ice age science report*

Pretend you're a scientist who just found a frozen baby mammoth.

Write a short "field report" on the next pages, including:

- Where you found it
- What it looked like
- What clues it might reveal
- What name you'd give it
- What question you'd want to answer

_____

_____

_____

_____

_____

_____

_____

_____

_____

_____

_____

_____

_____

_____

_____

_____

_____

bonus!

Draw your discovery scene and label the parts—fur, trunk, tusks, feet, and icy surroundings.

Photo credit: Mammuthus primigenius (baby woolly mammoth) (Late Pleistocene, 42 ka; Yamal Peninsula, Siberia, Russia) 1" by James St. John is licensed under CC BY 2.0.

7
what mammoths teach us

WOOLLY MAMMOTHS MAY BE GONE, but their story didn't end in the Ice Age.

Today, scientists are still learning from their bones, tusks, DNA, and frozen remains. Every discovery helps us understand not just mammoths—but the world they lived in, the changes they faced, and the choices we face today.

When you study the past, you also learn something important about the future.

from ice to insight

What do mammoths actually teach us? More than you might think. Their story reveals:

- **Climate matters.** Mammoths thrived in a cold world—but when the planet warmed too fast, they couldn't adapt quickly enough.
- **Humans make an impact.** Early people didn't mean to wipe out mammoths—but overhunting, along with other pressures, made a difference.
- **Nature is connected.** When mammoths vanished, whole ecosystems changed. Big animals often help shape the world around them.
- **Survival takes teamwork.** Mammoths relied on herds, memory, and cooperation. Just like humans do.

Their extinction wasn't caused by one thing. It was many changes happening at once—a powerful reminder that survival is never guaranteed.

Then vs. Now

Let's compare two moments in time.

Then: Ice Age

Mammoths roamed grassy plains

Ice sheets covered much of Earth

Humans hunted with spears

Nature shaped the future

Now: Today

Elephants roam forests & savannas

Glaciers are melting around the world

Humans change the world with cities, cars, and climate change

Now we shape the future

The question is: What do we do with that power?

saving modern giants

Woolly mammoths are gone—but their cousins are still here. Today's elephants face serious threats:

- **Poaching** (illegal hunting for ivory)
- **Habitat loss** (as forests disappear)
- **Climate change** (affecting food and water)
- **Conflict with humans** (in crowded areas)

If we don't protect them, they could go extinct too. That would mean losing a living link to the mammoth—and a key part of Earth's natural balance.

try this: *adopt a living cousin*

Choose one of these modern elephant species for your own research project:

- **African Savannah Elephant**
- **African Forest Elephant**
- **Asian Elephant**

Find out:

- Where it lives
- What it eats
- What dangers it faces
- What people are doing to help
- What you can do to make a difference

Draw a picture of your elephant in its habitat. Give it a name— and a voice. What would it say if it could speak to the world?

the big picture

The woolly mammoth story is more than a tale about a giant, shaggy animal from long ago. It's a reminder that:

- The planet changes
- Species come and go
- People play a big role
- The choices we make now **matter**

You don't need to be a scientist to care. You just need to be curious, brave, and ready to ask good questions—like the ones that helped us meet the mammoth in the first place.

Even though the mammoth's footsteps have faded... its story still marches on.

8
bringing mammoths back to life

A genetically engineered "woolly mouse"

WHAT IF WE could bring the woolly mammoth back?

Not just in movies or museums—but alive, breathing, and trumpeting across the frozen tundra once again. It sounds like science fiction, but scientists are working on it right now.

Welcome to the world of "de-extinction."

the big idea

Scientists discovered that woolly mammoth DNA is 99.6% identical to Asian elephant DNA. That tiny 0.4% difference is what made mammoths woolly, cold-adapted giants instead of tropical forest dwellers.

The plan: take an Asian elephant and give it just enough mammoth genes to make it look and act like a mammoth. A company called Colossal Biosciences thinks they can do it—and they're aiming for the first "mammoth" calves by 2027 or 2028.

how it would work

Creating a new mammoth is like being the world's most careful editor, changing just the right parts to tell a different story.

The process:

1. **Find the Right Genes** — Scientists identify genes that gave mammoths shaggy coats, curved tusks, and dome-shaped heads
2. **Edit Elephant DNA** — Using gene-editing tools, they replace elephant genes with mammoth versions
3. **Create an Embryo** — The edited DNA goes into an elephant egg
4. **Find a Surrogate** — An Asian or African elephant would carry the embryo
5. **Birth** — If it works, the world meets its first "mammoth" in thousands of years

practice round: woolly mice

Before trying with elephants, scientists started small. In 2025, they created "woolly mice" with mammoth traits like thick, golden-brown fur and cold tolerance.

These mice prove the approach works. "This is really validation that what we have in mind is really going to work," says Dr. Beth Shapiro from Colossal Biosciences.

why bring back mammoths?

Scientists think mammoth herds could help fight climate change. When mammoths roamed the tundra, they trampled snow and kept the ground frozen. This trapped greenhouse gases underground. Without them, that frozen ground melts and releases carbon into the air.

not exactly the same

These new "mammoths" wouldn't be identical to originals. They'd be elephant-mammoth hybrids with less than 1% true mammoth genes.

Think of it like giving a modern car the look of a 1950s classic—it might appear vintage, but underneath it's still modern technology.

Brain Boost: What Is De-Extinction?

"De-extinction" means creating a new animal that looks and acts very similar to an extinct species, using genetic engineering. It's like making a really good copy that could fill the same role in nature.

the great debate

Not everyone thinks this is a good idea.

Supporters say:

- Could help fight climate change
- Would restore important ecosystems
- Technology could help save endangered species

Critics say:

- Money should go to living species needing help now
- These hybrid animals might not survive in the wild
- We don't know how they'd affect ecosystems

what happens next?

If successful, the first mammoth-like calves could be born soon. They'd start in controlled environments, then maybe small herds would be released into protected Arctic areas.

It's one of science's most ambitious projects—bringing back a species gone for 4,000 years.

The question isn't just whether we *can* bring back the mammoth. It's whether we *should*.

try this: *design your de-extinction project*

Pick an extinct animal you'd like to see again. Research when and why it went extinct, its closest living relative, and whether bringing it back would help or harm today's world.

Present your findings: Should we try to bring your animal back?

What do you think? Should we focus on saving existing species, or also try to bring back what we've lost? The mammoth's story might just be beginning.

bonus fun!

mammoth mix-up

Unscramble these woolly words.

1. HETRD = _____
2. KUST = _____
3. CEFROTPRASM = _____
4. RUKNT = _____
5. DEXTNITC = _____
6. GEARSS = _____
7. MTHOMMA = _____

true or false?

Circle TRUE (T) or FALSE (F).

1. Mammoths lived on every continent on Earth. T / F
2. Some mammoths were found with hair and skin still attached. T / F

3. Mammoths were meat-eaters. T / F
4. The mammoth steppe was full of trees. T / F
5. Humans used mammoth bones to build shelters. T / F

quiz: what's your ice age superpower?

Circle one answer from each question to find your inner mammoth trait!

1. What would you pack for the Ice Age?

A. A super-warm coat

B. A big map and a sharp memory

C. A shovel and some snacks

2. What's your favorite herd activity?

A. Staying close to family

B. Finding new paths

C. Digging for food

3. If danger shows up, you...

A. Trumpet and rally your friends

B. Remember a safer place

C. Use your tusks to defend

Color the Woolly Mammoth!

glossary

Atlatl — A special tool used by ancient humans to throw spears farther and faster. Like a prehistoric slingshot for hunting.

Bachelor group — A small group of male animals (like mammoths or elephants) that live together after leaving their family herd.

Climate change — A big shift in Earth's weather patterns over time. Warming temperatures after the Ice Age helped lead to the mammoth's extinction.

Extinct — When a species has no living members left. Woolly mammoths are extinct, but their relatives (like elephants) still live today.

Fossil — The preserved remains or traces of an ancient organism. Fossils can include bones, teeth, tusks, or even footprints.

Herbivore — An animal that eats only plants. Mammoths were herbivores who loved grass, shrubs, and twigs.

Herd — A group of animals that live, move, and protect each other together. Mammoths, like modern elephants, traveled in herds.

Ice Age — A time in Earth's history when large parts of the planet were covered in ice and snow. The last Ice Age ended about 12,000 years ago.

Ivory — A hard material found in the tusks of elephants and mammoths. It was often carved into tools or art by early humans.

Matriarch — The oldest and wisest female in a herd, who leads the group and remembers where to find food and safety.

Mammoth Steppe — A massive grassland that stretched across Europe, Asia, and North America during the Ice Age. It was the perfect home for mammoths.

Permafrost — Ground that stays frozen year-round. In places like Siberia, it preserved entire frozen mammoths for thousands of years.

Predator — An animal that hunts and eats other animals. Mammoths had to watch out for Ice Age predators like cave lions and wolves.

Preserved — Protected from decay or damage over time. Some mammoths were preserved in ice so well that their fur, skin, and even stomachs survived.

Species — A group of living things that are similar and can have babies together. Mammoths were a species closely related to elephants.

Trunk — The long, flexible nose of an elephant or mammoth, used for smelling, grabbing, drinking, and trumpeting.

Tusk — A long, curved tooth made of ivory. Mammoths used their tusks to dig, fight, and protect themselves.

answer key

Answer Key on page 50.

mammoth mix-up

1. HERD
2. TUSK
3. PERMAFROST
4. TRUNK
5. EXTINCT
6. GRASS
7. MAMMOTH

true or false

1. FALSE (They didn't live in South America or Australia.)
2. TRUE
3. FALSE (They were herbivores.)
4. FALSE (The mammoth steppe was mostly grassland.)
5. TRUE

quiz: what's your ice age superpower?

———

Herd Protector (Mostly A's)

You're the guardian of the group! Like the matriarchs and adult females in a mammoth herd, you keep the little ones safe, trumpet alarms when danger is near, and stay close to family. Your strength isn't just in size—it's in loyalty and courage. When predators lurk or storms hit, everyone looks to you for protection and reassurance.

Fun Fact: Fossil evidence and elephant studies suggest mammoth herds were led by females who organized defense and surrounded calves when threatened.

———

Memory Keeper (Mostly B's)

You're the navigator and problem-solver! Just like today's elephants, mammoths likely relied on their oldest females to remember safe routes, seasonal grazing grounds, and water sources. Your sharp memory and sense of direction guide the herd through blizzards, across rivers, and back to familiar feeding spots year after year.

Fun Fact: Elephants use long-term memory to survive droughts and migrations. Scientists believe mammoths used the same skills to cross the vast "mammoth steppe."

———

Tusk Champion (Mostly C's)

You're built for action! With tusks that can shovel snow, dig for hidden grasses, and fend off rivals, you're the power player of the Ice Age. Like the big bull mammoths, you stand your ground in battles, clear paths through ice, and use your strength to keep the herd moving forward.

Fun Fact: Mammoth tusks—often over 12 feet long—show wear marks from digging and signs of clashes between males. They were survival tools as well as weapons.

more fossil files

Unearth the Giants of the Past!

Step back in time and meet the amazing giants that once ruled the Earth! *The Fossil Files* is a kid-friendly educational series that digs up the most amazing prehistoric predators and Ice Age icons and brings them roaring back to life—one fossil at a time.

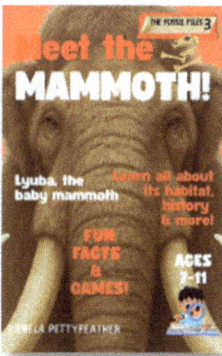

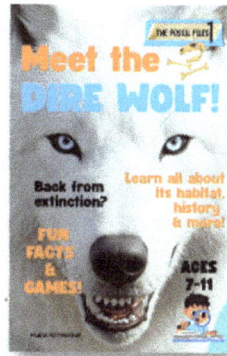

———

who's your ice age twin?

Find out which Ice Age creature you're most like!

Take this fun quiz and unlock your free *Fossil Files Activity Pack!*
Scan the QR code above (or tap the image below)!

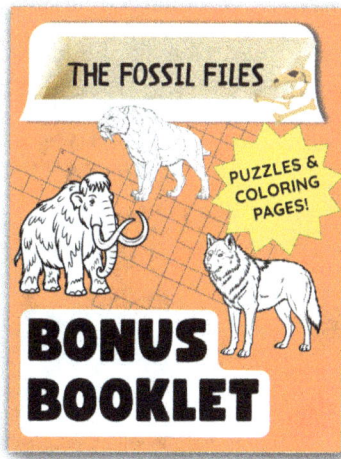

books for curious & clever kids

global explorations. bold questions. weird holidays.

Welcome to the *Books for Curious & Clever Kids series*—a collection of lively nonfiction books that bring world traditions, history, and culture to life for kids ages 8–12!

These books are designed for the endlessly curious and the slightly-too-clever kids who always ask: "Wait… Why do we do that?" "Where did that come from?" "Do other people do that, too?!"

Each book in the series explores a major holiday's surprising history and global story with Kid-friendly activities.

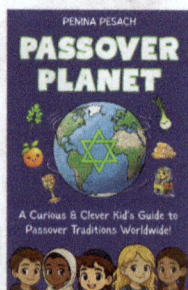

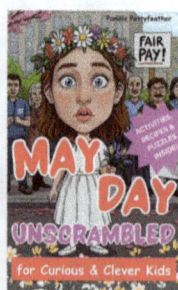

easter unscrambled for curious & clever kids

Why do we hunt for eggs on Easter? Where did the Easter Bunny come from? And what do ancient gods and chocolate bunnies have in common? *Easter Unscrambled* is the perfect blend of fascinating facts and hands-on fun — ideal for kids ages 7-12 who love puzzles, history, and surprises. This interactive book takes readers on a 4,000-year adventure through the origins of Easter — from ancient gods and springtime festivals to Easter egg hunts and chocolate bunnies.

———

halloween unscrambled for curious & clever kids

Why do we carve pumpkins? Where did trick-or-treating really come from? And what does a 2,000-year-old Celtic festival have to do with your Halloween candy? *Halloween Unscrambled* is a fascinating journey through history that reveals the surprising true story of how Halloween became the holiday we celebrate today.

From ancient Celtic bonfires to Mexican Día de los Muertos altars, discover how Halloween traditions traveled around the globe and evolved over centuries!

passover planet: a curious & clever kids' guide to passover traditions worldwide!

Frogs. Pharaohs. Pancakes. Passover Like You've Never Seen It Before! Get ready to travel through **3,000 years of Passover celebrations**—from **ancient Egypt to modern space stations**, from Moroccan mimouna feasts to Ecuadorian chocolate charoset! In *Passover Planet*, kids will **1)** Discover global Passover traditions from **Ethiopia, Persia, Poland, Iraq, South America** & more **2)** Learn about cool customs like scallion fights and Afikoman ransoms **3)** Explore **Seders with matzah pizza**, spicy charoset, and teff flatbread…

---

may day unscrambled for curious & clever kids

Why do people dance around poles in spring… while others wave protest signs? From leafy legends to labor marches, *May Day Unscrambled* unravels one of the world's most surprising holidays! Kids will travel through time and across the globe as they explore: 1) **Ancient spring festivals** like Beltane and Walpurgisnacht **2) British traditions like maypole dancing and the May Queen** 3) The rise of **International Workers' Day** and the fight for fair labor and 4) Global May Day celebrations in **France, Finland, Hawaii**, and more!

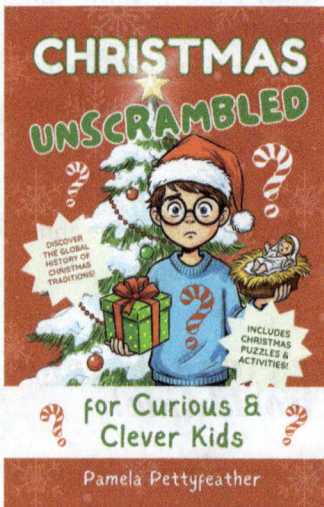

---

christmas unscrambled for curious & clever kids

Why do we hang stockings, bake gingerbread, and wait for a man in a red suit? *Christmas Unscrambled* peeks behind the traditions to reveal the history, legends, and global customs that shaped the holiday. Kids will explore: 1) Solstice festivals like Saturnalia and Yule 2)The origins of Santa Claus and gift-giving 3) Holiday foods, songs, and symbols around the world 4) Christmas celebrations from Europe, Africa, Asia, and the Americas

www.ingramcontent.com/pod-product-compliance
Lightning Source LLC
Chambersburg PA
CBHW050605280326
41933CB00011B/1991